Year 7 Mathematics Practice Papers

About the book

There are 4, year 7 Mathematics papers & answers in this book. These are 2 sets of papers 1 (non-calculator) & 2 (calculator) written as practice papers for end of year 7 Mathematics Examinations in June 2020. Papers are mainly focusing on topics covered by most schools in year 7 mathematics syllabuses in The United Kingdom. However, you may still use this book as a practice for other syllabuses for 11 to 12 year olds.

All the questions in this book are written by the author and they are new questions written purely to help and experience the students to prepare and test themselves for the upcoming end of year mathematics exams.

Answers are included in this book. If you need to check your solutions, I advise you to ask your school mathematics teacher or your private mathematics tutor to mark your answers.

There are 2 sections to this book A & B. Each section contains 2 papers. The first paper of each section is a non-calculator paper & the second paper of each section is a calculator paper.

Year 7 Mathematics Practice Papers

(Year 7 Mock Exams)

for 11 to 12 year olds

4 mock papers including answers

By Dilan Wimalasena

Contents

	Page
Section A	7
Section A Paper 1	9
Section A Paper 2	17
Section B	25
Section B Paper 1	27
Section B Paper 2	37
Answers	45

Section A

Year 7

Mathematics

Practice Paper A1

June 2020

Calculator is not allowed

**Time allowed
1 hour
Total 100 marks**

Write answers in the space provided

Year 7 Mathematics Practice Papers

1. Write the following fractions in their simplest form

i) $\frac{12}{18} = \frac{2}{3}$ ii) $\frac{6}{8} = \frac{3}{4}$

(2 marks each)
(total 4 marks)

2. John leaves home at 7.48am every morning to go to school. It takes him 25 minutes to walk to the school. At what time does John reach the school?

08:13 am

(3 marks)

3. Convert the following measurements into metres.

i) 200cm = 2 meters

(2 marks)

ii) 2km = 2000 meters

(2 marks)
(total 4 marks)

4. Here are the numbers 1 to 10.

$$1, 2, 3, 4, 5, 6, 7, 8, 9, 10$$

i) write down a multiple of 3. 6

(2 marks)

ii) write down 3 prime numbers 1, 2, 7

(2 marks)

iii) write down 3 even numbers 2, 4, 6

(3 marks)
(total 7 marks)

11

5. Work out the following

i) $-2 + 5 =$ 3

(2 marks)

ii) $-2 - 5 =$ −7

(2 marks)

iii) $2 - 5 =$ 3

(2 marks)

iv) $2 + 5 =$ 7

(1 mark)
(total 7 marks)

6. Complete the missing terms of the sequence

24, 21, ~~19~~ 18, 15, 12, ~~16~~ 15, ~~30~~ 21 ...

(3 marks)

7. Points A, B, C have coordinates $A(1,2), B(3,7)$ & $C(4,2)$.

i) Plot the points A, B & C on the grid below. Label your x & y axes clearly.

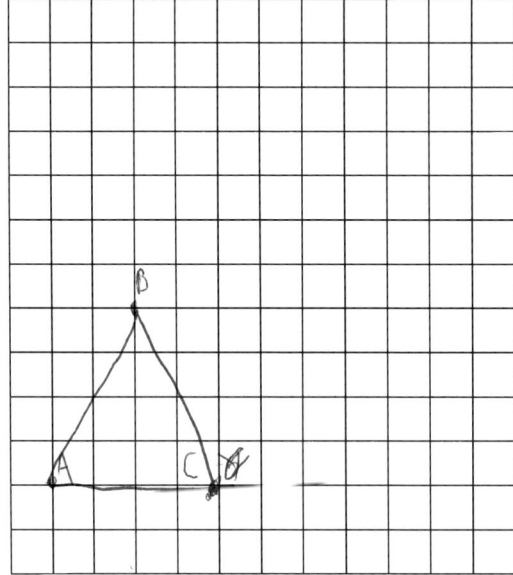

(3 marks)

ii) Draw the triangle ABC.

(1 mark)
(total 4 marks)

8. Solve the following equations

i) $a + 8 = 19$

(2 marks)

ii) $a - 3 = 15$

(2 marks)

iii) $2a = 18$

(2 marks)

iv) $\dfrac{a}{3} = 4$

(2 marks)
(total 8 marks)

9. John had £120. He gave $\dfrac{1}{4}$ of it to his father. How much does he have now?

£90

(4 marks)

10. Work out the value of x in each diagram and write a reason for each question.

i)

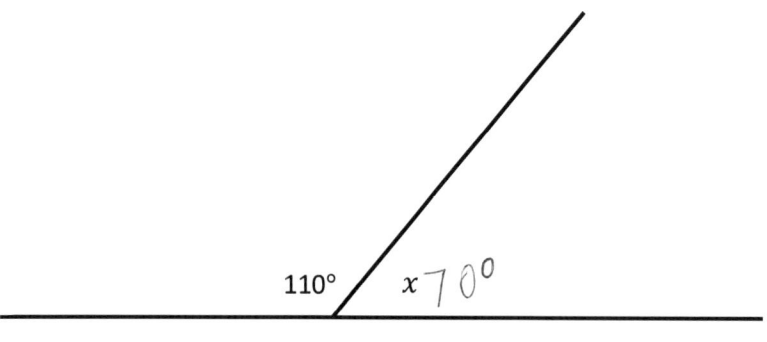

(3 marks)

ii)

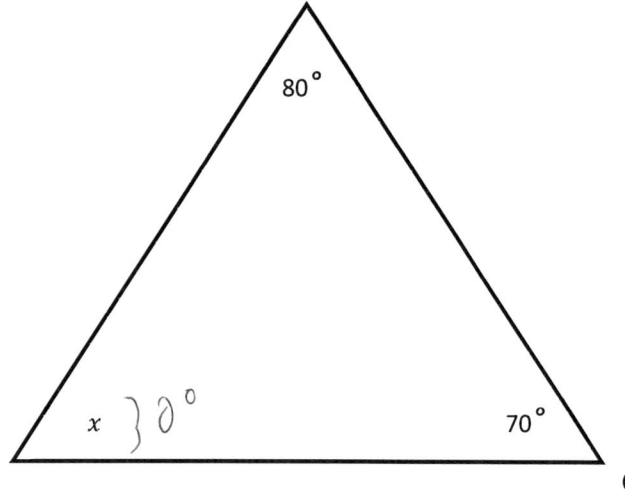

(4 marks)

iii)

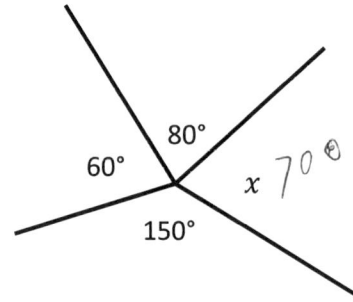

(3 marks)
(total 10 marks)

11. Work out the following

i) $2(3 + 5) - 4 =$ ~~8~~ 6

(3 marks)

ii) $5 \times 4 - 10 \div 2 =$ 15

(3 marks)
(total 6 marks)

12. Write following percentages as fractions

i) 50% $\frac{1}{2}$

(2 marks)

ii) 75%   ~~$\frac{3}{4}$~~

(3 marks)

iii) 40%   ~~$\frac{2}{5}$~~

(2 marks)

iv) 5% $\frac{5}{100}$

(3 marks)
(total 10 marks)

13. Find Highest Common Multiple (HCF) of 40 & 50. 10

(4 marks)

14. Find Lowest Common Multiple of 8 & 12. 4

(6 marks)

15. Simplify the following fully

i) $5a - 2a =$

(2 marks)

ii) $4y + 3y - 5y =$

(2 marks)

iii) $2(2x + 3) =$

(2 marks)

iv) $5(7y - 1) =$

(2 marks)

v) $6a + 5b - 2a - 7b =$

(3 marks)
(total 11 marks)

16. Calculate perimeters of shapes below.
*please note shapes are not drawn accurately.

i)

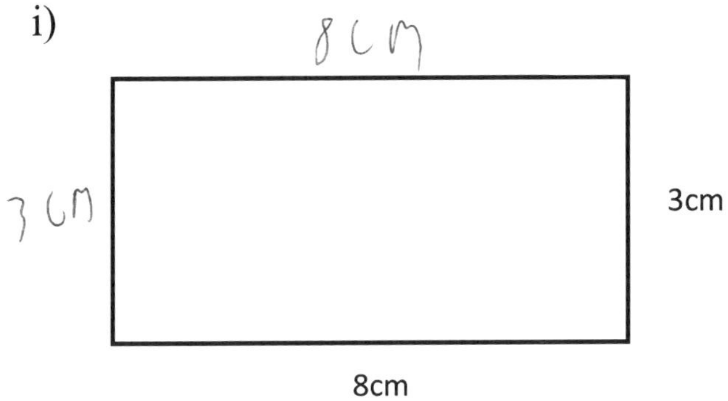
8cm
3cm
3cm
8cm

22cm

(3 marks)

ii)

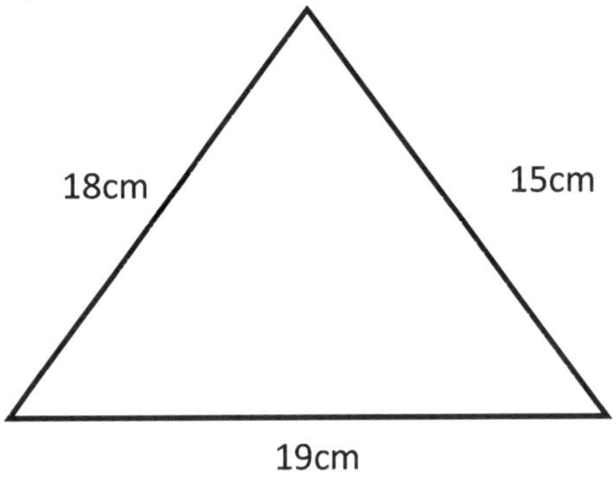
18cm
15cm
19cm

55cm

(3 marks)

iii)

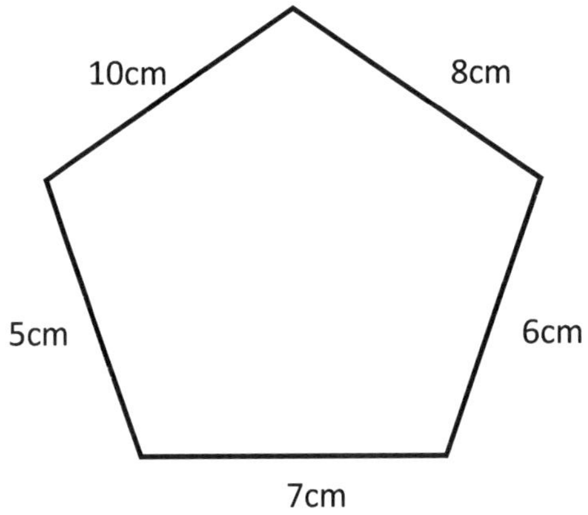
10cm
8cm
5cm
6cm
7cm

36cm

(3 marks)
(total 9 marks)

Total for paper: 100 marks

End

Year 7

Mathematics

Practice Paper A2

June 2020

Calculator is allowed

**Time allowed
1 hour
Total 100 marks**

Write answers in the space provided

1. Joanna sells cakes for £7.50 each. She sold 18 cakes. Calculate the total she received from sales.

(3 marks)

2. i) Write the following list of numbers in ascending order

$$123, 118, 121, 103, 199, 200, 100$$

(3 marks)

ii) Write the following list of numbers in descending order

$$78, 51, 68, 41, 56, 57, 81$$

(4 marks)
(total 7 marks)

3. Work out the following
i) 28×13

(3 marks)

ii) 46×15

(3marks)

iii) 1.28×1.31

(3 marks)
(total 9 marks)

4. Write down the missing terms

$i)$ $25, ..., 31, 34, ..., 40$

(2 marks)

$ii)$ $1.5, 2.5, ..., ..., 5.5$

(2 marks)

$iii)$ $1.75, 1.65, ..., ..., 1.35$

(2 marks)
(total 6 marks)

5. Work out the following

$i)$ $\dfrac{4}{5} - \dfrac{1}{4} =$

(3 marks)

$ii)$ $\dfrac{2}{3} \times \dfrac{4}{5} =$

(2 marks)

$iii)$ $\dfrac{1}{3} \div \dfrac{4}{7} =$

(2 marks)
(total 7 marks)

6. Find the missing number in each case

$i)$ $2 \times .... + 9 = 19$

(2 marks)

$ii)$ $4 \times 6 - .... = 17$

(2 marks)

$iii)$ $3 \times .... = 12 \times 2$

(2 marks)
(total 6 marks)

7. Work out areas of following shapes.

i)

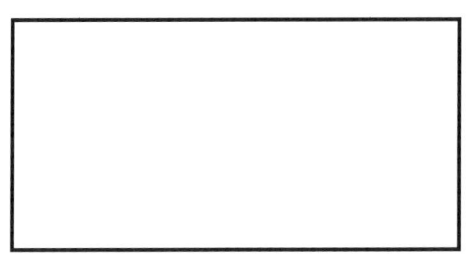

(3 marks)

ii)

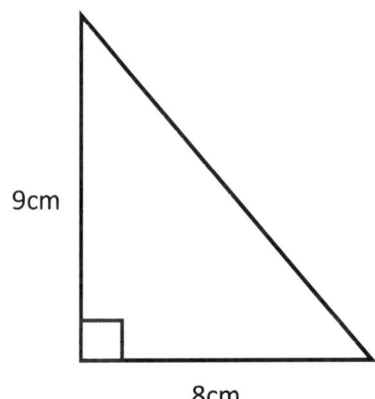

(3 marks)

iii)

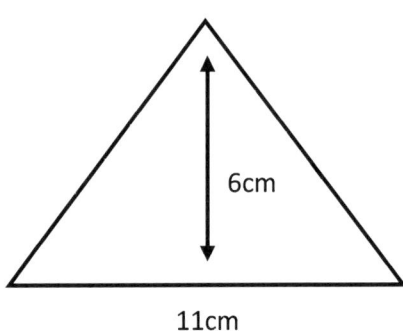

(3 marks)
(total 9 marks)

8. Convert the following fractions into improper fractions.

i) $2\frac{1}{3}$

(3 marks)

ii) $4\frac{2}{5}$

(3 marks)

iii) $5\frac{3}{7}$

(3 marks)
(total 9 marks)

9. Convert the following fraction into mixed fractions

i) $\frac{12}{5}$

(3 marks)

ii) $\frac{3}{2}$

(3 marks)

iii) $\frac{10}{3}$

(3 marks)
(total 9 marks)

10. Work out the value of angle x.

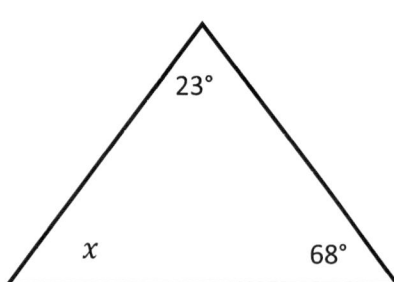

(5 marks)

11. Calculate the volume of the cuboid.

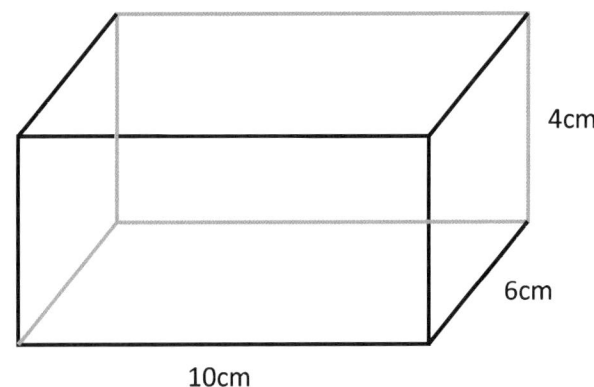

(3 marks)

12. Write down all prime numbers between 10 and 30.

(3 marks)

13. Represent the following data in a bar chart. (state your axes clearly.)

Weights(kg)	Number of people
10-20	3
20-30	5
30-40	6
40-50	4

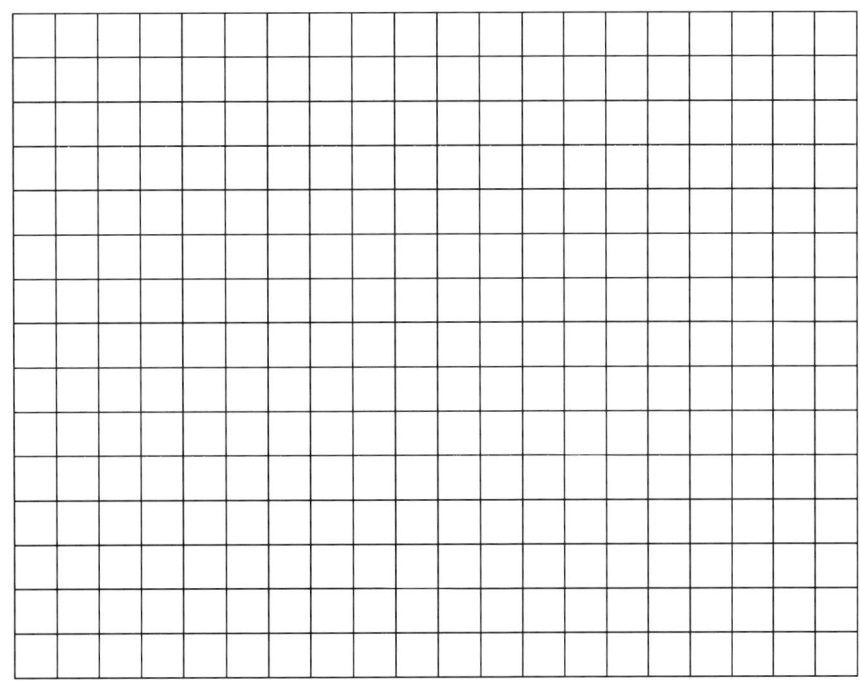

(5 marks)

14. Calculate 25% of £84.

(3 marks)

15. Calculate the mean value of following list of numbers.

$$23, 19, 46, 31, 55$$

Write your answer to one decimal place.

(4 marks)

16. Convert the following fractions into percentages.

i) $\dfrac{3}{4}$

(3 marks)

ii) $\dfrac{1}{4}$

(3 marks)

iii) $\dfrac{2}{5}$

(3 marks)

iv) $\dfrac{3}{8}$

(3 marks)
(total 12 marks)

Total for paper: 100 marks

End

Section B

Year 7

Mathematics

Practice Paper B1

June 2020

Calculator is not allowed

**Time allowed
1 hour
Total 100 marks**

Write answers in the space provided

1. Work out the following

i) $\dfrac{2}{5} + \dfrac{1}{4} =$

(3 marks)

ii) $\dfrac{3}{4} - \dfrac{1}{3} =$

(3 marks)
(total 6 marks)

2. i) Sam is 12 years old. His father is 3 times his age. What is his fathers' age?

(3 marks)

ii) Nelly is 16 years old. Her sister is 5 years younger than her. What is her sisters' age?

(3 marks)
(total 6 marks)

3. Work out the following

i) 234×17

(4 marks)

ii) 315×68

(4 marks)
(total 8 marks)

4. Work out the highest common multiple (HCF) and lowest common multiple (LCM) of 24 & 36.

(6 marks)

5. Work out the following

i) $2^3 =$

(2 marks)

ii) $3^2 =$

(2 marks)

iii) $10^3 =$

(3 marks)
(total 7 marks)

6. Write the following fractions as percentages.

i) $\frac{4}{5}$

(3 marks)

ii) 0.28

(3 marks)

iii) $£2 \; out \; of \; £4$

(4 marks)
(total 10marks)

7. Calculate 1.2×3.4

(4 marks)

8. How many seconds are there in an hour?

(3 marks)

9. $a = 5, b = -2, c = 3.$
Work out
i) $2a + 3c$

(3 marks)

ii) $4c + 5b$

(4 marks)
(total 7 marks)

10. Find the value of x.
(diagram not accurately drawn)

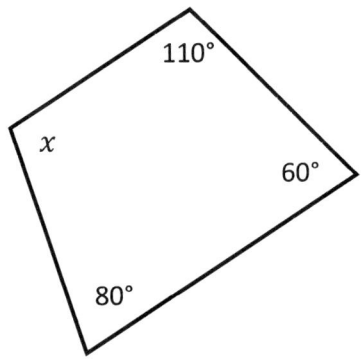

(4 marks)

11. A quadrilateral has vertices $P(1,3)$, $Q(1,7)$, $R(5,3)$, $S(5,7)$

i) Plot the points P, Q, R, S on the grid below.
(clearly label your x & y axes.)

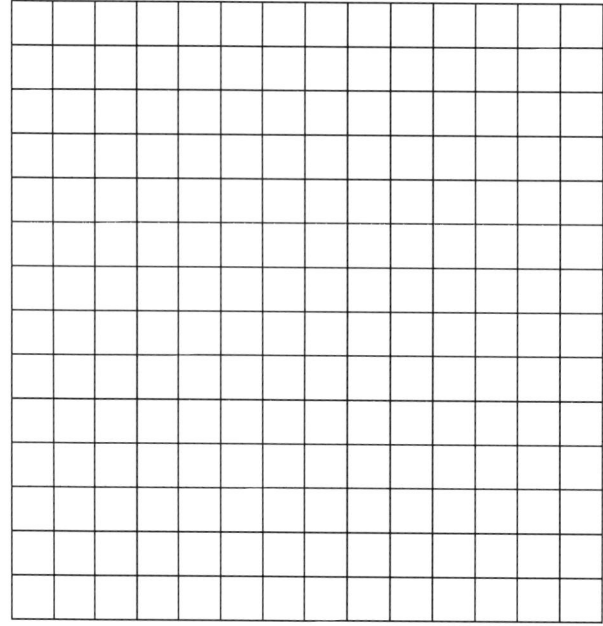

(4 marks)

ii) Complete the quadrilateral by connecting the points.

(2 marks)

iii) What type of a quadrilateral is $PQRS$?

(2 marks)
(total 8 marks)

12. Work out the following
i) $2.3 \times 10 =$

(2 marks)

ii) $4.21 \times 100 =$

(2 marks)

iii) $23.3 \times 100 =$

(2 marks)

iv) $2.33 \times 10 =$

(2 marks)
(total 8 marks)

Year 7 Mathematics Practice Papers

13. Solve the following equations
$i)\ x + 4 = 10$

(3 marks)

$ii)\ y - 2 = 11$

(3 marks)

$iii)\ 2y + 3 = 13$

(4 marks)

$iv)\ 5a - 7 = 13$

(4 marks)
(total 14 marks)

14. Round the following numbers to the nearest 10.
i) 29

(2 marks)

ii) 111

(2 marks)

iii) 199

(2 marks)

iv) 235.5

(3 marks)
(total 9 marks)

Total for paper: 100 marks

End

Year 7

Mathematics

Practice Paper B2

June 2020

Calculator is allowed

**Time allowed
1 hour
Total 100 marks**

Write answers in the space provided

1.

```
        _____
       |            |
       |            | 3cm
       |_____|
          7.5cm
```

i) Calculate the perimeter of rectangle.

(4 marks)

ii) Calculate the area of rectangle.

(4 marks)
(total 8 marks)

2. Work out the following

$i)\ 2 \times (3 + 2) - 7 =$

(3 marks)

$ii)\ 4 \div (3 - 1) + 10 =$

(3 marks)
(total 6 marks)

3. i) Sally tossed a coin. What is the probability of getting tails?

(2 marks)

ii) John rolled a dice. What is the probability of getting number 4?

(2 marks)
(total 4 marks)

4. Find the missing angle x in each case.
(write a reason for each question)

i)

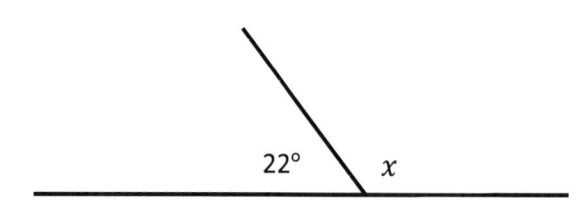

(3 marks)

ii)

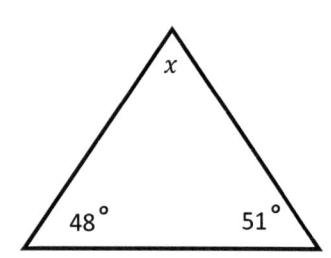

(4 marks)

iii)

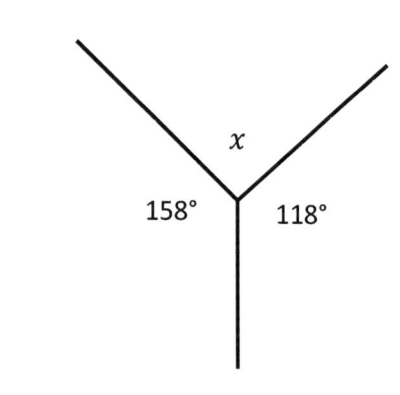

(4 marks)

iv)

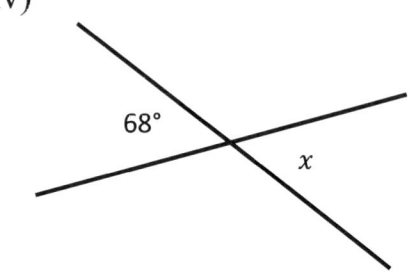

(3 marks)
(total 14 marks)

5. Work out the highest common factor (HCF) of the following
i) 20 & 30

(3 marks)

ii) 12 & 18

(3 marks)

iii) 48 & 60

(4 marks)
(total 10 marks)

6. John and Mary shared £80 in the ratio 1:3.
Work out the amount received by Mary?

(5 marks)

7. Calculate the following

i) 2^4

(2 marks)

ii) 3^3

(2 marks)

iii) $2^4 + 3^3$

(2 marks)
(total 6 marks)

8. Solve the following equations.

i) $5a = 2a + 12$

(4 marks)

ii) $3y = 35 - 4y$

(4 marks)
(total 8 marks)

9. Simplify the following fully

i) $5x + 2x - 3x =$

(3 marks)

ii) $2(4x + 5y) =$

(3 marks)

iii) $3(4y + 5) + 2(2y + 3) =$

(4 marks)
(total 10 marks)

10. Kevin played 11 games of cricket and below are his scores.

$$22, 41, 22, 18, 22, 7, 11, 44, 50, 9, 17$$

i) Work out the median score.

(3 marks)

ii) Work out the mean score.

(3 marks)

iii) Work out the mode.

(2 marks)
(total 8 marks)

11. In a group of students, 15 people like English, 6 like science & 9 like mathematics.

Represent this on a pie chart.

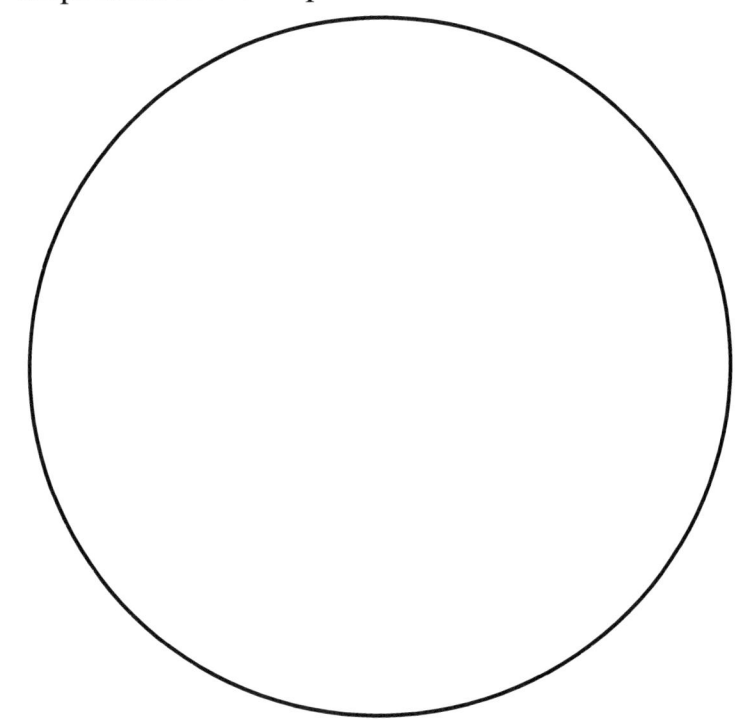

(5 marks)

12. Calculate the volume of cuboid.

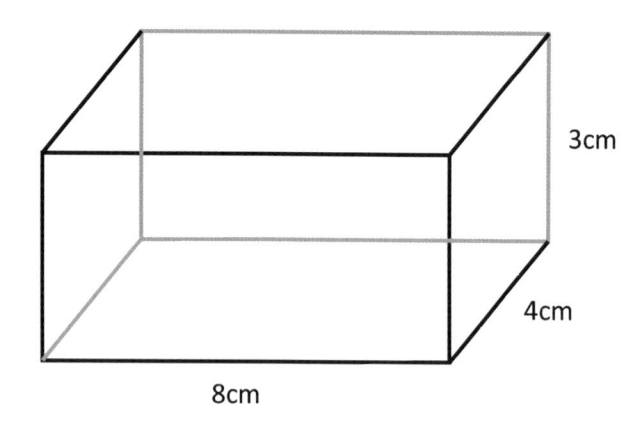

[Cuboid with dimensions 8cm, 4cm, 3cm]

(4 marks)

13. Work out the following

i) $3\frac{1}{2} - 1\frac{2}{3} =$

(4 marks)

ii) $3\frac{1}{3} \times 1\frac{4}{5} =$

(4 marks)
(total 8 marks)

Total for paper: 100 marks

End

Answers

Year 7 Mathematics Practice Papers

Paper A1	Paper A2
1. i) $\frac{2}{3}$, ii) $\frac{3}{4}$	1. £135
2. 8.13am	2. i) 100, 103, 118, 121, 123, 199, 200 ii) 81, 78, 68, 57, 56, 51, 41
3. i) $2m$, ii) $2000m$	3. i) 364, ii) 690, iii) 1.6768
4. i) 3, 6 or 9 ii) any three from 2, 3, 5 or 7. iii) any three from 2, 4, 6, 8, 10	4. i) 28, 37 ii) 3.5, 4.5 iii) 1.55, 1.45
5. i) 3, ii) -7, iii) -3, iv) 7	5. i) $\frac{11}{20}$, ii) $\frac{8}{15}$, iii) 7/12
6. 18, 9 & 6	6. i) 5, ii) 7, iii) 8
7. i) plot points correctly ii) Join ABC to complete the triangle.	7. i) $28cm^2$, ii) $36cm^2$, iii) $33cm^2$
8. i) 11, ii) 18, iii) 9, iv) 12	8. i) $\frac{7}{3}$, ii) $\frac{22}{5}$, iii) $\frac{38}{7}$
9. £90	9. i) $2\frac{2}{5}$, ii) $1\frac{1}{2}$, iii) $3\frac{1}{3}$
10. i) 70° (angles on a line add up to 180°) ii) 30° (angles in a triangle add up to 180°) iii) 70° (angles at a point add up to 360°)	10. 89°
11. i) 12, ii) 15	11. $240cm^3$
12. i) $\frac{1}{2}$, ii) $\frac{3}{4}$, iii) $\frac{2}{5}$, iv) $\frac{1}{20}$	12. 11, 13, 17, 19, 23, 29
13. HCF = 10	13. Draw a bar chart with correct heights.
14. LCM = 24	14. £21
15. i) $3a$, ii) $2y$, iii) $4x + 6$, iv) $35y - 5$, v) $4a - 2b$	15. 34.8
16. i) 22cm, ii) 52cm, iii) 36cm	16. i) 75%, ii) 25%, iii) 40%, iv) 37.5%

Year 7 Mathematics Practice Papers

Paper B1	Paper B2
1. $i) \frac{13}{20}, ii) \frac{5}{12}$ 2. $i)$ 36, $ii)$ 11 3. $i)$ 3978, $ii)$ 21420 4. HCF = 12, LCM = 72 5. $i)$ 8, $ii)$ 9, $iii)$ 1000 6. $i)$ 80%, $ii)$ 28%, $iii)$ 50% 7. 4.08 8. 3600 seconds 9. $i)$ 19, $ii)$ 2 10. 110° 11. i) plot the points correctly ii) join points to draw the quadrilateral iii) Square 12. i) 23, ii) 421, iii) 2330, iv) 23.3 13. $i) x = 6, ii) y = 13, iii) y = 5, iv) a = 4$ 14. i) 30, ii) 110, iii) 200, iv) 240	1. $i)$ $21cm, ii)$ $22.5cm^2$ 2. i) 3, ii) 12 3. $i) \frac{1}{2}, ii) \frac{1}{6}$ 4. $i)$ $158°$ $(angles\ on\ a\ line)$ $ii)$ $81°$ $(angles\ in\ a\ triangle)$ $iii)$ $84°$ $(angles\ at\ a\ point)$ $iv)$ $68°$ $(vertically\ opposite\ angles)$ 5. i) 10, ii) 6, iii) 12 6. £60 7. i) 6, ii) 27, iii) 43 8. $i) a = 4, ii) y = 5$ 9. $i) 4x, ii) 8x + 10y, iii) 16y + 21$ 10. $i)\ median = 22$ $ii)\ mean = 23.9$ $iii)\ mode = 22$ 11. $English\ 180°, Science\ 72°, Mathematics\ 108°$ 12. $96cm^3$ 13. $i)$ $1\frac{5}{6}, ii)$ 6

47

Printed in Great Britain
by Amazon